Janine Blinko
Illustrations by Alison Dexter
Photographs by Zul Mukhida

Contents

A

Introduction

The long-term aim of early years education in mathematics is to produce adults who are confident, competent and creative in all aspects of the subject. To achieve this, the short-term aim is to present young children with a '...broad, balanced and purposeful curriculum' (*Looking at Children's Learning*, SCAA, July 1997).

The National Numeracy Strategy *Framework for teaching mathematics* (DfEE, 1999) and *Early Learning Goals* (QCA, 1999) offer a structure within which teachers can make this happen. Some key, common elements to these documents provide the framework for this book:

- teaching is based on identified learning objectives or goals;
- mathematics sessions are structured and focused, and maintain a good pace;
- assessment includes informed observation and oral questioning;
- pupils are questioned effectively;
- the early stages of learning are oral;
- where possible, mathematical ideas are supported at home in a fun and interesting way.

Shape, space and measure offers a curriculum for nursery and reception aged children that gives them the opportunity to do mathematics for themselves, to think about and develop their ideas, and to discuss them. When the children go on to Year 1, they will have learned how to tackle the subject with confidence, and have a sound foundation on which to build.

The National Numeracy Strategy and the early learning goals emphasise that practical activity, discussion and focused teaching are the triple key to understanding mathematics. This has long been acknowledged, and is arguably more true in the area of space, shape and measure than in any other area of mathematics.

Providing the right equipment is not enough on its own. Children learn mathematics best if it is presented in an enjoyable way and discussed with skilled and informed questioning from the teacher.

At this early stage, learning is directed by the children themselves; the teacher's role is to interject with questions that move children's thinking forward in the course of their play. For example, if a child makes a model of a person in Plasticine, you might ask, 'Now can you make a taller person?' or 'Now can you make a person with longer legs?' When children are at the stage where they can respond appropriately to such questions, it is time to structure their learning more tightly with clear learning goals. The activities in this book aim to build on children's early 'free' experiences and focus their thinking on the mathematics in which they are involved.

How to use this book

The first unit, which focuses on shape and space, looks closely at circles, squares and triangles – the most familiar two-dimensional shapes that children will encounter. Most of these activities can be adapted to focus on other polygons, if appropriate.

The second unit focuses on measurement and, in particular, the language of measurement. There is a huge array of words associated with measurement and children need plenty of practice at both hearing the words and using them in a range of contexts. Wherever possible, they should be encouraged to repeat phrases back to the teacher or to a friend to help them fix their understanding.

The third unit looks at themes that can be used to explore several concepts. Often, a single theme, such as a box, can lead to a wide range of mathematical ideas, and it is a shame not to use them when an opportunity arises.

Differentiation

It is not expected that all children will experience all the activities in a particular chapter. Within each activity, there are extension ideas for children who have demonstrated sound understanding in the main task. There are also suggestions for varying the activities to help those children who need to consolidate their ideas by doing the same thing in a slightly different context.

Assessing the children's learning

The main method of assessment is observing how children cope with the activities. The assessment section in each chapter suggests ideas for focused questions. There are also descriptions of responses that children have given as they have carried out these activities in the classroom, which may also be helpful for assessment purposes.

Other quick ideas

'Other quick ideas' are activities for a spare five minutes. Requiring little preparation, they are an important way of presenting children with new ideas as well as consolidating their understanding, and can quickly become a regular and enjoyable part of the day.

Involving parents

Finally, each chapter includes suggestions for several simple activities that parents can carry out at home with the children, in recognition of the important part that parents can play.

The table below shows the connections between activities in this book, early learning goals for mathematical development and key objectives in the *Framework for teaching mathematics* for Reception. Some activities link with more than one goal or key objective; the teacher/trainer's questions and subsequent discussion with the children will provide the focus.

Early Learning Goals and Key Objectives	Unit: Activities	
Use language such as 'circle' or 'bigger' to describe the shape and size of solids and flat shapes;	**Introducing shape:** all the activities **Circles:** Lids **Triangles:** Skipping ropes **Squares:** Colourful metre square	**Boxes:** all the activities **Balls:** Paint balls; Marbles; Rolling **Lines:** Stringy lines; Line up; Car tracks
begin to name flat shapes such as circle, triangle, square and rectangle; [key objective only]	**Introducing shape:** Peeping; Fabric shapes **Circles:** Walk a circle; I spy circles	**Triangles:** Cutting triangles; Dotty triangles **Squares:** Overlapping squares **Balls:** Paint balls
use everyday words to describe position;	**Reflections:** Mirrors; Body shapes	**Where?:** Pathways and mazes; Treasure hunt; Who is next to whom?
use language such as 'more' or 'less', 'heavier' or 'lighter' to compare two numbers or quantities;	**Comparing:** all the activities **How long? How tall?:** all the activities **Fill it up:** Trickle lines; Stones; Holes **How heavy?:** all the activities	**When?:** all the activities **Lines:** Paint lines **Bags:** Cotton wool balls
use developing mathematical ideas and methods to solve practical problems.	**Introducing shape:** Shape detectives **Circles:** Find the circle **Triangles:** Dotty triangles; Body triangles **Squares:** Joining squares **Reflections:** Copy you, copy me; Magic pictures **Where?:** I'm thinking of a thing; Treasure hunt	**Comparing:** Where in the room? **How long? How tall?:** Lengths **Fill it up:** Overflow **How heavy?:** Is it heavy or is it light? **When?:** Pegs; Birthdays in order **Balls:** Tell me about it **Bags:** Make your own bag; Sort; Dressing-up bags

Introducing shape

Intended learning

To distinguish between two- and three-dimensional shapes; to recognise the common properties of shapes; to name and recognise two- and three-dimensional shapes.

Introduction

Many older children are confused about the properties of shapes. Much of their confusion arises from their early experiences of being introduced to regular shapes with little mention of irregular shapes. A regular shape is one that has all its sides and angles equal. Using this definition, rectangles, parallelograms, and irregular hexagons, pentagons, octagons, and so on are not regular shapes.

There is also a great deal of specific language associated with shapes. There is no reason why children cannot be introduced to and learn correct terms from the start, for example, vertices and angle. These terms and others are listed in the glossary.

This chapter focuses on activities that make connections between the properties of a range of flat (two-dimensional) and/or solid (three-dimensional) shapes.

Key vocabulary

straight, curved, vertex, corner, edge, side, face

Fabric shapes

Group size:
up to six children.

You will need:
a collection of fabric shapes (either purpose-made or ready-made items such as a triangular scarf, a doily, a tea-towel, a handkerchief).

The activity
Introduce the activity by hiding one of the fabric shapes in your pocket or in your hands. Show the children a small portion of the edge or a corner of the shape. Ask them what shape they think the item will be. Gradually reveal more and more of the piece of fabric, inviting the children to offer suggestions as to the shape. Finally smooth it out flat in front of the children and ask them to identify its shape.

Invite the children to take turns to hide the fabric shapes, challenging each other to identify them. Use every opportunity to emphasise the appropriate language ('How did you know it was a triangle?' 'How many corners does it have?' 'Are its sides straight?' 'What other shapes have straight sides?' 'Do you know any shapes that don't have straight sides?').

Extension and differentiation
- The process can be made simpler by using shapes made from plain fabric, with no geometrical designs that might prove confusing.

- If children are struggling to identify the shapes, comment on the properties that can be observed as you reveal the shape ('I can't see any corners and the side seems to be curved...').

- Make some large shapes from paper and fold them. Ask the children to guess what the final shape will be as you unfold them slowly.

Printing with shapes

Group size:

four to six children.

You will need:

large sheets of paper; paint; a selection of three-dimensional shapes to print with; an identical set of shapes.

The activity

Let the children use the shapes to print with. Discourage them from using the shapes to spread paint across the paper, but rather focus on printing shapes and encouraging them to identify the shapes they are producing.

When the pictures are dry, look at the shapes together. Let the children handle the spare set of three-dimensional shapes. Can they guess which three-dimensional shapes were used to print which pictures? Can the children name each of the two-dimensional shapes they have made? What shapes did they expect? Were there any surprises (for example, they might expect a square-based pyramid to print only triangles)? Did anyone print with a sphere? What happened?

Extension and differentiation

- If children find using more than one shape confusing, give them just one shape each. Look at all the pictures together and help the group to match up the correct two- and three-dimensional shapes.

- Ask them to print using every face of each shape just once. Which three-dimensional shape has the most faces?

The children challenged each other to identify the fabric shapes.

Peeping

Group size:
up to ten children.

You will need:
a set of cards showing two-dimensional shapes; a screen of some sort (for example, a book or a thick piece of card).

The activity
Introduce the activity by showing the children all the cards and discussing the shapes on them. Then explain to the children that you are going to hide the shapes behind a screen. The shapes are going to peep out a little at a time and the children must try to guess what they are. Choose one of the shapes, and reveal it bit by bit from behind the screen, pausing occasionally, until the children can guess which shape it is. Continue until the whole shape is revealed.

Once the children are fairly confident with this, let them take turns to hide the shapes.

Extension and differentiation
- To make the game easier, provide two identical sets of shapes, and display one for the children to look at as they try to guess which one is peeping.

- Repeat the activity but, this time, use large three-dimensional shapes (boxes work well). This will help to focus the children on specific attributes of a shape.

Shape detectives

Group size:
up to ten children.

You will need:
a collection of two- and three-dimensional shapes; a tray or table.

The activity
Place the shapes on a tray or table in front of the children and ask them to look at the shapes carefully. (The number of shapes you choose will depend on the children's experience and level of understanding.)

Think of one of the shapes, without telling the children what it is. Explain that you have chosen a shape and you are going to give them clues so that they can be shape detectives and find out which one it is. If you choose a cube, you might give the following clues.

- It has eight vertices/corners.
- It has six faces.
- It has square faces.

Once the children have played this a few times, let them take turns to choose a shape and think up clues themselves.

Extension and differentiation
- To make the game simpler, use fewer and less complex shapes.

- Repeat the game, but tell the children that they must ask you questions to get their clues ('Does it have more than six edges?' 'Is it a two- or three-dimensional shape?' 'Does it have a curved edge?').

Assessment

- Can the children identify and name various two- and three-dimensional shapes?
- Are they beginning to use the vocabulary of shape in an appropriate way?
- Can they link two- and three-dimensional shapes?
- Are they beginning to understand and recognise the properties of different shapes?

Evidence of the children's learning

There was a great deal of excitement in the *Peeping* game, but it soon became obvious that some children were really drawing on their knowledge of which shapes were in the set, using phrases like, 'There were two with those corners,' and 'It might be the...'. In *Shape detectives,* the children's use of the necessary language developed with each game. For the first few games, they had trouble retaining all the properties that were discussed, but soon began to talk and argue with each other about what shape it could or could not be.

Other quick ideas

- Let the children take turns to walk around an imaginary shape, then ask the others to guess what shape it was.
- Cut some shapes out of sandpaper and stick them on to pieces of card. Put the cards in a bag and let the children take turns to put their hands in and try to guess what the shapes are.
- Ask the children to use interlocking cubes or other, similar equipment to make three-dimensional shapes.
- Provide a collection of two- and three-dimensional shapes, then ask the children to take turns to find a shape that does not have a particular property (you could ask them to find a shape which does not have curved sides or a shape which is not a triangle).

Involving parents

Ask parents to look for some fabric shapes at home, and let their children bring them in to show everyone. Suggest a number of shape-related activities that could be undertaken at home, such as making shape biscuits, using Plasticine to make three-dimensional shapes, and collecting old boxes of all shapes and sizes to make models.

Circles

Intended learning

To identify a circle from a collection of shapes, including other round shapes, such as oval, ellipse, egg shape; to begin to know the properties of a circle.

Introduction

This and the following two chapters explore common two-dimensional shapes, namely circles, squares and triangles. Reference is also made to other polygons.

Circles are more difficult to define mathematically than any other polygon. Polygons are generally defined in terms of their vertices (corners) and sides, but a circle is either defined as having one side and no corners or as having an infinite number of sides and corners.

Young children can also find the language associated with circles rather confusing, particularly when the term is used colloquially, not precisely. For example, 'Sit in a circle' actually means 'Imagine a circle drawn on the floor and sit around its side/s'. There are also occasions when we say 'round' in conjunction with 'circle' (as in 'a round circle') despite the fact that a circle can never be anything but round!

Key vocabulary

circle, circular, round, shape, curved, centre

Walk a circle

Group size:

up to eight children.

You will need:

a skipping rope.

The activity

Ask all the children to stand facing you, and then to turn all the way around to face you again. You may well need to demonstrate this the first time, turning around on the spot.

Invite two children to move into a clear space in front of the others. Give the children one end each of a skipping rope. Tell the first child to stand still and hold the rope firmly, and the second child to walk around the first child. Explain that they must both hold tightly on to the rope and keep it stretched between them. Tell the rest of the group to watch carefully.

Let other children have a turn, and discuss what happens. Encourage questions and responses, for example, 'Where is the first child?' (*At the centre of the circle.*) 'When is the second child closest to the centre?' (*Never!*) 'When is the second child furthest from the centre?' (*Never!*) 'How do you know?' (*The rope is always the same distance from the centre.*)

Extension and differentiation

● Help children to understand what is happening by letting them record the circle pathway. Get the group to either place a cube on every footprint as the second child walks the circle or chalk the circle on the floor.

● Let the children work in pairs using a shorter piece of string. Help them to tie a pencil on to one end, hold the other end still, and draw a circle on a piece of paper.

Find the circle

Group size:

up to six children.

You will need:

old, clean socks; a selection of flat or nearly flat objects of different shapes, including circles (for example, shapes cut from card, a coin, a postcard, etc.); a piece of string; clothes pegs.

The activity

Tie the string across the room or between two chairs to make a 'washing line'. Secretly put one shape in each sock and peg them along the line. (The number of socks you use depends on how many children there are and how long you want the game to last. However, you should try to use several examples of circles.)

Introduce the activity by discussing circles and maybe showing the children some circular objects. Explain that you have hidden some circular objects inside the socks, and that they will need to find them by feeling. Let the children take turns to feel the socks and try to find one with a circle inside. Explain that they may feel as many as they choose, but they may only peep inside one. When they find a circle in a sock they can hold on to it until the end of the game. The winner is the child who has correctly identified the most circles.

Extension and differentiation

- Support children by letting them feel an assortment of circular objects before the activity. Encourage them to close their eyes and talk about what they can feel.

- Put shapes in a bag, and let the children take turns to reach inside to try to find a circle. Let them keep the circle if they do find one.

- Rather than letting the children feel lots of socks, let them only have one try. This makes the game more difficult as they have to remember what other children have done.

- Ask the children to find different shapes.

Revealing the circle hidden in a sock

I spy circles

Group size:
up to eight children.

You will need:
a range of circular objects, or objects featuring circles, placed around the room (cups, drinking straws, chair legs, round table tops, jars, lids, bottles, cans, and so on).

The activity
This game is much the same as the traditional I spy game. However, rather than 'spying' things which begin with a particular letter, the children should only spy things made up of circles! Begin by giving the children a clue about where to look, for example, 'I spy with my little eye, something near the table which has a circle on it.'

Children often get quite good at this game, and it provides an ideal opportunity to discuss the differences between shapes which are circles, and those which are not quite circles, such as ovals or octagons.

Extension and differentiation
● Support children by pointing out a number of circular objects before the game and talking about their properties. Introduce the word 'circular' when playing the game.

● Give the children different shapes to look for ('I spy with my little eye something on the wall which has a rectangular shape').

● Dispense with giving the children a location and just give them a shape to look for.

Lids

Group size:
four to six children.

You will need:
a selection of circular lids from jars and bottles; trays of paint; paper.

The activity
Show the children how to print with the lids by dipping just the rims carefully into the paint trays. Once they can do this, let them design their own pictures made up of circles. Explain that the circles must be clear; discourage them from dragging the lids across the paper.

When the children have finished and the paintings are dry, allow them to look at each other's designs. Encourage discussion by asking questions ('Has anyone made one where the circles overlap?' 'Has anyone made one where the circles don't overlap?' 'What colour are the largest circles?' 'What about the smallest?'). Draw attention to any pictures that represent things, for example circles that make flowers.

Extension and differentiation

● Challenge the children to make a picture of an animal (such as a caterpillar) or a person using different types of lids.

● Develop the activity by giving the children several lids of different sizes and showing them how to print concentric circles.

● Ask the children to make pictures where all the circles must overlap with another circle.

● If children find the activity confusing, let them use just one lid to make a specific picture (such as oranges and apples).

Evidence of the children's learning

In *Find the circle*, the children voluntarily used the language of shape to talk about what they were feeling. There were some shapes in the socks which were also curved (ellipses and crescents) and some which were near circles, like jam pot lids. The children became very particular about what makes a circle, and developed the phrase 'completely round' to describe it and differentiate it from the others! One child referred back to the *Walk a circle* game that we had played the previous day as a way of explaining what is special about a circle.

Assessment

● Can the children recognise a circle when presented with a collection of regular and irregular shapes, as in *Find the circle* or *I spy circles*?

● Can the children explain why a shape is or is not a circle?

Other quick ideas

● Give the children different-sized circles and a small plastic mirror. Let them experiment with the different shapes they can make by holding the mirror across the circles in different places.

● Roll out some Plasticine into long thin sausages. Use them to make circles, or stick them round the edge of circles drawn on paper or card.

● Provide a set of three-dimensional shapes made of card or foam. Ask the children to find any circular faces of the shapes and to print with them.

● Give the children sticky circle shapes to make pictures with.

Involving parents

Ask parents to let their children draw around some circular-shaped objects at home (plates, egg cups, mugs, etc.). How many different sizes can they find? They could also show children how to use paper and wax crayons to make rubbings of circular shapes (screw heads, jar lids, coins, etc.).

Triangles

Intended learning

To identify a triangle from a collection of polygons; to begin to understand the properties of a triangle; to use appropriate language to describe shapes.

Introduction

A triangle can be defined as a two-dimensional closed shape with three straight sides. In later years one of the main misunderstandings children have about triangles is that 'proper' ones must look like this:

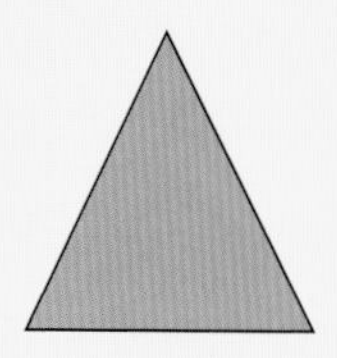

Consequently they will describe any other triangle in relation to this one. Thus they will refer to the following triangles as an upside-down triangle and a flat triangle.

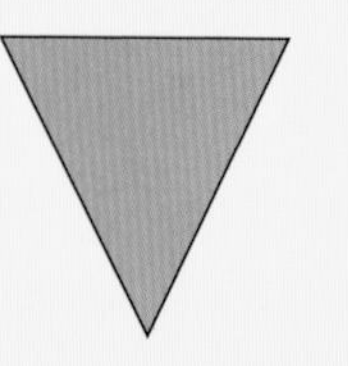

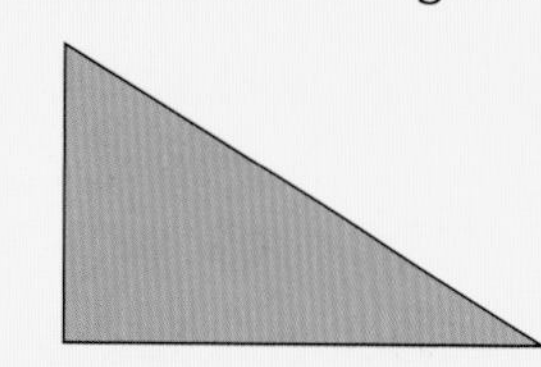

The activities in this chapter present various triangles in different orientations and contexts to encourage children to develop a general concept of the shape.

Key vocabulary

triangle, triangular, corner, vertex, side

Cutting triangles

Group size:

up to five children.

You will need:

a selection of triangular objects and triangles cut from a variety of materials; sheets of paper of different sizes and colours; a pair of scissors for each child.

The activity

Introduce the activity by showing the children some triangles. The greater the variety of shape and size, the better. Discuss with the children the name of the shapes. Lead them to understand that they are all triangles. Count the sides and the vertices. Explain that the shapes are all triangles because they all have three straight sides and three vertices.

Ask the children each to choose a piece of paper and to cut a straight line across it from corner to corner. Then invite them to cut again and again to make lots of triangles. Stress that the cuts must be as straight as possible. Discuss what they have made: 'Are all the shapes triangles?' 'How many triangles did you make?' 'Could you make any more?' 'Did you make any other shapes?' 'How do you know they are triangles?' 'Who has made the largest?' 'Who has made the smallest?' 'Has anyone made two the same?'

Extension and differentiation

- Demonstrate the cutting process and count the sides of the resulting shape together, or cut the triangles for them.
- Sort the resulting triangles into size order.
- Use the triangles to make a picture.

Dotty triangles

Group size:

up to six children.

You will need:

a sheet of paper; a pencil.

The activity

Before the activity, use the pencil to cover the paper with irregularly spaced dots.

Show the children the piece of paper and point out three dots that you have 'chosen'. Join the three dots with lines to make a triangle.

Pass the sheet round the group for each child to have a turn at joining three dots to make a triangle. Discuss the triangles the children have made ('Who has made the thinnest?' 'Who has made the largest?' 'Which triangle has the sharpest corners?').

Extension and differentiation

- If children are having trouble seeing the triangles in a page of dots, repeat the activity using fewer dots more regularly spaced.
- Encourage the children to overlap triangles on the same sheet.
- Ask the children to make sure that each of the triangles joins on to another one.
- Let the children have their own sheet with dots on and make their own set of triangles.

The children found out that joining three dots together produces a triangle.

Skipping ropes

Group size:
four to six children.

You will need:
plenty of skipping ropes.

The activity
Introduce the activity by reminding the children what a triangle is, and how many sides and vertices it has.

Tie the ends of each skipping rope together. Ask the children to work in pairs to lay the rope on the ground to make a triangle. Talk about the triangles they have made and discuss whether they are long and thin or short and fat. Talk about the different orientations. Then ask the children to change their shape to make a different triangle. How does the new triangle differ from the previous one?

Extension and differentiation
- Support children by giving them pictures of triangles or triangular objects to refer to as they work.
- Join three skipping ropes together to make a giant triangle.
- Ask the children to make a skipping rope triangle and fill it with shoes, books or some other object. How many fit inside exactly? Will more fit if the triangle is arranged differently?
- Make other shapes with the ropes.

Body triangles

Group size:
up to six children.

You will need:
no special equipment.

The activity
Introduce the activity by sitting in front of the group and asking them to copy your movements. Then use your finger to draw a triangle in the air. Ask the children to try to identify the shape. You may need to repeat it several times.

Then try using other body parts, such as a foot or an elbow, to draw triangles in the air, again asking the children to copy you.

Finally let the children take turns to choose body parts with which to draw a triangle in the air. Ask the rest of the group to copy them.

Extension and differentiation
- Help the children to make, rather than draw, triangle shapes with parts of their bodies. They could make a triangle using thumbs and index fingers, or they could work with a partner to make a triangle from three feet.
- Let the children try to draw pictures to record their favourite 'body triangles'.
- Try the same activity again, but focus on different shapes.

Assessment

- Can the children recognise a triangle?

- Do they understand that a three-sided shape with straight edges is a triangle, whatever its size or orientation?

- Can the children explain why a shape is or is not a triangle?

Evidence of the children's learning

When the children were cutting triangles, they were fascinated by the fact that they could cut many different types of triangles just with one cut, and that provided the shape had three straight sides it was still a triangle. They also began to ask questions about the other shapes that they made (often accidentally!), and learnt the word quadrilateral to describe a shape with four straight sides.

Other quick ideas

- Give the children different-sized triangles cut from paper or card and a small plastic mirror. Let them experiment with the different shapes they can make by holding the mirror across the triangles in different places.

- Let the children use a ruler to try to draw some triangles.

- Give the children some identical paper triangles, and ask them to stick them together so they do not overlap and there are no spaces between them. (This provides an introduction to the concept of tessellation.) Can this be done with all triangles?

- Make triangles with straws and pipe cleaners.

- Experiment with sticky paper triangles to make repeating patterns.

Involving parents

Ask parents to let their children use anything appropriate to make triangles (cutlery, straws, long blades of grass, and so on). They could be asked to help children fold paper triangles.

Squares

Intended learning

To identify a square from a collection of four-sided shapes, such as rectangle, parallelogram, irregular quadrilaterals (four-sided shapes), and so on; to begin to understand the properties of a square.

Introduction

A square is a two-dimensional closed shape with four straight sides of equal length and four right-angled corners. One of the main misunderstandings children have about squares is that they are only squares if they are oriented like this: but not like this:

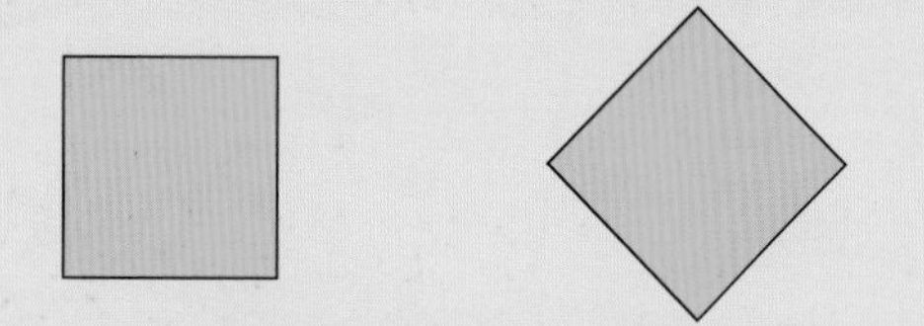

If a square is rotated, many children feel that it is more correctly called a 'diamond', which is not the correct mathematical name for any polygon. The activities that follow give children the opportunity to see, draw and discuss squares of all sizes. In all the activities, encourage them to check whether shapes are squares by finding a way to measure the four sides. For example, they could cut a piece of string to the same length as one of the sides and measure the others against it, or they might fold the shape diagonally.

Key vocabulary

square, square corner, straight, side, corner, vertex

Colourful metre square

Group size:

as many children as possible.

You will need:

10-centimetre square pieces of paper; felt-tipped pens; crayons; large sheet of display paper; glue.

The activity

Depending on the number of children in the group, this activity could take place over a number of sessions.

Explain to the children that they are each going to decorate a square of paper, and that all the squares will be joined together to make a big, special-sized square. Give the children one square each and ask them to decorate it however they choose, using the crayons or felt-tipped pens. When they have finished their squares, ask the children to place them side by side on the table or on the floor until there are ten in a row. Then start a new row of ten squares directly underneath, and so on, until a metre square has been formed.

Discuss the shape with the children as it develops: 'Is it a big square yet?' 'How can we check?' 'Will it be a square when the current row is completed?' 'Will it be complete if everyone makes another square?'

Stick the finished squares in position on backing paper and display it with a suitable caption. Make the square a focal point for discussion for the children over the next few days, encouraging them to count the squares, to try to find their own squares, and so on. It will be a big achievement when it is complete!

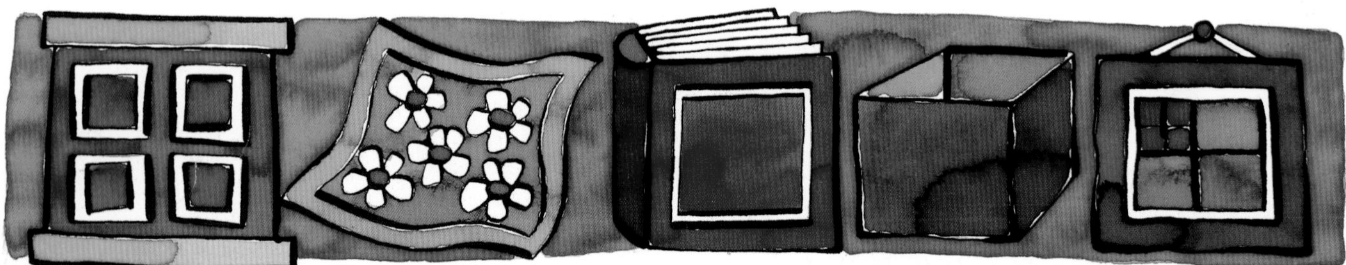

Extension and differentiation

- If children need help visualising the metre square that they are aiming for, it is helpful to draw the outline for them and discuss the fact that the big square will have the same number of small squares along each side.

- If the children wish, they can decorate another square or take one home to decorate.

- Give the children larger squares (for example, 25 cm x 25 cm) to decorate with paint or with printing.

- One or two children could be invited to decorate a metre square with paint applied with large domestic paintbrushes.

Joining squares

Group size:

up to six children.

You will need:

lots of paper squares of the same size.

The activity

Introduce the activity by showing the children the squares and ensuring that they all know the name of the shape. Give them two squares each and ask them to place them together to make different shapes. Ask them to find as many different ways as possible. Ask them to count how many sides there are on each new shape they make.

Colouring in the 10 cm squares to make the giant square

Note: the names of the shapes are not important; what matters is that the children try to make different shapes with different numbers of sides.

Extension and differentiation

● Support children by letting them use craft putty to hold their shapes in place while you count the sides with them.

● Let the children glue their squares in place and use them for a display.

● Specify the number of sides of a shape and ask the children to try to make it.

● Use triangles or rectangles instead of squares for the activity.

Overlapping squares

Group size:

up to six children.

You will need:

lots of different-sized squares (preferably in various colours) cut from old greetings cards and other sources.

The activity

Invite children in the group to take turns to find the correct size of square to make arrangements of squares that get smaller and smaller, or bigger and bigger.

Extension and differentiation

● If necessary, reduce the number of squares to make the activity simpler.

● Try the same activity with different shapes.

● The children could glue their squares in place to make a display.

Assessment

● Can the children recognise a square?

● Can they identify a square from a selection of quadrilaterals?

● Can they explain why a shape is or is not a square?

Evidence of the children's learning

In most of the activities, the checking of the squares became a main focus. The children particularly enjoyed the folding method to check that two adjacent sides were the same length. They were not really aware that they had only checked one pair of adjacent sides, but were happy that if the two halves of the square sat on top of each other, then they had a true square.

There was also plenty of incidental learning about squares when the children made the metre square. They preferred not to fill the metre in rows,.but to make growing squares made up firstly of one small square, then four squares, then nine squares, and so on.

Other quick ideas

● Provide 2 x 2 squares cut from 5 centimetre squared paper, and ask the children to colour them using just two colours. Let them use the squares to make chequered flags, by joining them together to make larger squares and then gluing them on to a flagpole made out of a drinking straw.

● Challenge the children to draw a square using a ruler.

● Make pictures from sticky paper squares (square people, dogs, houses, and so on) to make a display entitled 'Square land'.

● Make square hats for a 'Square party' with square pretend food (sandwiches, cakes, etc.), on square plates cut from card, and square party invitations.

Involving parents

Invite parents to help their children find square shapes around the house, such as the seat of a dining room chair, a box lid, paving slabs, windows, etc. The children could then draw round some of the squares and cut them out.

Reflections

Intended learning

To identify reflective patterns in pictures; to create shapes which have reflective symmetry.

Introduction

Children have a natural feeling for symmetry. They often draw people symmetrically, despite the fact that we seldom stand symmetrically; the children observe and use the symmetry anyway! The activities in this chapter build on this innate understanding by encouraging the children to look for symmetry in nature, by emphasising their own and others' body symmetry, and by letting them create symmetry of their own. Children's understanding of symmetry is not an exact one, so an emphasis needs to be placed on checking whether each 'side' of the reflection is exactly the same as the other. It may be appropriate to use the word 'symmetry' to identify this property, but bear in mind that many children will not remember it.

Key vocabulary

reflection, reflect, symmetry, line, copy, the same, match

Mirrors

Group size:

up to six children.

You will need:

a plastic mirror for each child; some pictures of things that have 'natural symmetry' (flowers, people, leaves, etc.); pictures of shapes (both regular and irregular) that have 'constructed symmetry'.

The activity

Introduce the activity by showing the children how to hold the mirrors across the pictures to make new shapes. Explain that the picture they see in the mirror is called the reflection. Give them plenty of time to explore with their mirrors and discuss what they see. You may want to direct with some questions, such as: 'Can you make the picture look the same as it does on the paper?' 'Can you make it look like a different shape?' 'How can you tell that your picture is exactly the same as it was?' 'Can you make the shape or picture look longer/shorter/more pointed?'

Extension and differentiation

- Reinforce the concept of symmetry by letting the children take their mirrors around the room to find out if there are any classroom objects which have symmetry (a leaf, a saucer in the play house, etc.).

- Draw some 'half shapes', such as half of a triangle, and ask the children to use their mirrors to try to complete the shape.

- Repeat the activity but, this time, use shapes that are not symmetrical.

Body shapes

Group size:

up to 15 children.

You will need:

a fairly big space to work in, such as the hall or playground.

The activity

Explain to the children that they are going to make some shapes with their bodies. Give them a chance to experiment with moving their bodies into different shapes. Then ask them to try to hold themselves still in one particular shape.

Teach the children to make three symmetrical body shapes. Then invite the children to explore and create symmetrical body shapes of their own. Give them a chance to see the shapes that other children have made, and to try them out for themselves. Discuss with the children what it is that makes the shapes symmetrical. Lead them to understand that the shapes they are making should be the same on both sides if they are to be symmetrical.

Extension and differentiation

- Make a large mirror available so that the children can appreciate their own symmetrical body shapes.

- Choose four shapes for the whole group to learn, then make up a dance with them.

- Ask the children to draw a picture of themselves in one of their positions.

The children enjoyed making symmetrical body shapes.

Magic pictures

Group size:
four to six children.

You will need:
sheets of carbon paper; plain paper; pencils.

The activity
Prepare the paper for the children by putting the carbon sheet carbon-side down on the plain paper. Fold the plain paper in half so that the carbon sheet is folded backwards inside. Prepare enough for the children to have one each, plus one for demonstration purposes.

Demonstrate to the children how to draw a simple shape on the folded paper. Then open the paper out and remove the carbon paper. Show the children what has happened to the inside of the folded paper that was next to the carbon sheet. What do they notice about the picture that has appeared? Use and explain the term 'symmetrical' to the children as they describe the picture.

Give the children an opportunity to draw magic pictures for themselves.

Extension and differentiation
● Support children by providing mirrors so that they can see the reflective symmetry of the picture they have made.

● Encourage the children to predict the outcomes of the pictures they are drawing.

● Challenge the children to draw a picture that will show one of the following once it has been unfolded: one square, two triangles, two people.

Copy you, copy me

Group size:
up to ten children.

You will need:
pictures of butterflies with open wings; paper; coloured pencils or felt-tipped pens.

The activity
Introduce the activity by showing the children pictures of butterflies and discussing the symmetry of their open wings.

Ask the children to watch while you fold a sheet of paper in half. Open the sheet out flat again and point out the crease down the middle. Invite one of the children to draw a simple picture or shape on one side of the paper. Then copy their picture or shape on the other side of the fold for them.

Give each child a sheet of paper and help them to fold it carefully down the middle in the same way. Draw a simple picture or shape on the left-hand side of each child's paper. Discuss the pictures. Then ask the children to try to copy them on the other side of the fold.

Extension and differentiation
● If children find it difficult to draw a reflected image on the other half of the paper, provide them with mirrors and let them observe the reflected image before trying to draw it.

● Allow the children to work in pairs, each trying to copy the other's pictures.

● Provide cut-out butterfly shapes with patterns drawn on one wing only. Ask the children to copy the patterns exactly on to the other wing. (The complexity of the pattern will depend on the ability of the child.)

Evidence of the children's learning

The children loved using the mirrors and were fascinated with the different shapes they could make with them. They needed encouragement to check how precise they were being, but most of them became really particular about this work. Two children asked to use the mirrors to check the pictures that were made in *Copy you, copy me*, and were very happy to declare that some of them were 'not quite good!'

Assessment

- Can the children identify whether or not shapes are symmetrical?
- Can they create their own symmetries?

Other quick ideas

- Fold paper in half and cut out small shapes. Then try to match the shapes to the spaces.
- Paint a shape on one half of a sheet of paper. Fold the paper, press it down then unfold it to make a symmetrical print.
- Invite the children to collect things that are symmetrical (a leaf, a flower, a hat, etc.) for a symmetry display. Provide a mirror so that children can check the display for symmetry.

Involving parents

Ask parents to help their children find symmetrical things about the home (a chair, a mirror, a rug, patterns on carpets, curtains, wallpaper and clothes, and so on), and to look for symmetrical pictures in books. They could also cut open some fruit and vegetables, and look for symmetrical patterns inside them.

Where?

Intended learning

To use the language of position and movement; to talk about where things are in relation to themselves and to each other.

Introduction

There is a great deal of language that children need to understand to enable them to describe accurately where something is, both in terms of position and the route between two or more objects (over, under, forwards, backwards, etc.). When they first learn to walk, young children often go in a direct line to where they want to go, irrespective of what might be in the way! Later, they learn to walk around the objects that are in their path. This is the beginning of their understanding of where things are in relation to each other. The next stage is to develop an understanding of where they are in relation to other things and to appreciate how objects relate to each other. This section builds on this early appreciation of position, both in relation to the children themselves, and also to spatial relationships between other objects.

Key vocabulary

position, over, under, forwards, backwards, on, in, top, bottom, front, back, side, direction, across, up, down, towards, away from, turn, right, left, round, above, below, through, next to, in front of, behind, beneath, beside, between

Treasure hunt

Group size:

up to ten children.

You will need:

'treasure' (such as a gold-coloured box or a bag of gold-wrapped chocolate coins).

The activity

In advance of the activity, hide the treasure somewhere in the room.

Tell the children a story to set the scene (for example, Captain Blackbeard has hidden some treasure in the classroom and has left a number of clues to help find it again). Invite one child or a pair of children to find the treasure. Give them clues, using as many directional words as possible, for example, ask them to walk around the table, pass the sink, crawl under the sand tray, then look next to the door.

Repeat the activity with other pairs of children.

Extension and differentiation

- If children are having difficulty with the directional words, simplify the instructions and use the same words several times, for example, 'Crawl *under* the sand tray, then look *under* the cupboard.'

- Ask two children to go to another part of the room and cover their eyes. With the rest of the group, hide the treasure. Then ask the pair to return, and let the group describe the route to the treasure.

- Let the children work in pairs and hide the treasure for each other.

- Make a simple treasure map to follow.

The children were able to follow directions to find 'treasure' that had been hidden in the room.

Pathways and mazes

Group size:
up to five children.

You will need:
interlocking cubes; small toy vehicles; toy people or animals.

The activity
Introduce the activity by building pathways together using the bricks. Each path needs to be enclosed by two walls of bricks, to make a place for the vehicles, people or animals to pass along. Include corners in the path.

Let the children make the toys travel along the path. Talk with the children about what they have made and how they are using it ('How many corners are on the road?' 'Which car will you take along the road?' 'Where will you start?'). Encourage them to begin to use language of direction (forwards, backwards, round, left, right, and so on).

Extension and differentiation
- Children who are struggling with the activity can be helped by making the pathway simpler with fewer corners.

- Make the path more complicated to include some junctions and dead ends.

- Make a maze with Plasticine walls. Ask the children to find a route through the maze and describe it as they travel. Ask them to find and describe the longest and shortest routes. Invite them to direct a friend through the maze.

Who is next to whom?

Group size:
four to six children.

You will need:
table and chairs.

The activity
Sit the children around a table, making sure there are no gaps and no empty chairs. Discuss where everyone is sitting ('Who is facing Trevor?' 'Who is opposite you?' 'Who is in the corner?').

Then ask the children to take it in turns to say a sentence to describe where one person in the group is sitting in relation to another, encouraging them to use the correct language.

Extension and differentiation
- Reinforce the activity by inviting the children to change places. Then repeat the discussion.

- Ask the children to draw a 'map' of their positions around the table.

I'm thinking of a thing

Group size:
up to six children.

You will need:
no special equipment.

The activity
Ask one of the children to think of something in the room, such as a tap, the door or a light switch, and whisper to you what it is. Prompt the rest of the group to ask the child questions using positional language to try to discover what the object is: 'Is it near the door?' 'Is it next to the wall?' 'Is it in the sink?'

After several questions, give the group an opportunity to guess what the object is and where it is.

Extension and differentiation
- Simplify the game by giving clues, for example, 'I'm thinking of something that is next to the wall.'

● Invite the children to take it in turns to make up a sentence about where classroom objects are in relation to each other. ('The teddy is behind the pot.' 'The bottle is next to the big book.' 'The flower is in front of the egg cup.')

Assessment

● Can children use the language of direction appropriately?

● Can they use the language of position with accuracy?

● Do they respond appropriately to instructions which use the language of direction and positional vocabulary?

Evidence of the children's learning

The children became really animated when they talked about the mazes they had made, but needed encouragement to 'slow down' and describe them in terms of how many corners they had made, which were the longest routes, and which vehicles had the furthest to travel!

The other games quickly became everyday activities. Once the children understood them, we would play them in any spare moments and soon the children developed a broad base of words that they used when describing positions. They would also correct each other if the words were used inappropriately.

Other quick ideas

● Play 'Simon says' or 'Old Grady says' (put your hands behind your head, put your finger on your nose, etc.).

● Ask the children to direct you somewhere in the classroom.

● Invite the children to take turns to direct the whole class to the hall, to the playground, to the tree at the end of the field, etc.

Involving parents

Ask parents to collect old magazines or catalogues and encourage their children to look for examples of spatial relationships, such as things or people travelling along paths or roads; things inside other things; things over, under, round other things, and so on. Suggest that children might occasionally be allowed to direct the way home, or to the shops.

Comparing

Intended learning

To compare two or more objects; to identify which of two objects is faster, longer, heavier, higher, wider, lighter, narrower, etc.

Introduction

This chapter is the first of four chapters about the early development of measuring skills.

Learning how to measure plays a large part in the mathematics curriculum at this level and there are a number of stages in the development of the skills that children need: direct comparison of two objects (for example, holding two lengths of string next to each other and comparing their length); ordering two or more objects; measuring with non-standard measures; and seeing the need for and using standard units of measure.

The activities in this unit introduce comparison skills in the contexts of measuring length, weight, time and capacity.

Key vocabulary

light, wide, narrow, short, long, heavy, longer, heavier, higher, wider, lighter, narrower, faster, slower, more quickly, more slowly, pendulum

Pendulums

Group size:

four to six children.

You will need:

pieces of string cut to different lengths (but none longer than a child's arm); Plasticine.

The activity

Ask each child to choose one length of string. Help them to roll balls of Plasticine (roughly the size of marbles) and attach them to one end of the string. Then ask them to hold the other end. Point out how the weighted string now swings from side to side. Introduce the word 'pendulum'.

Discuss what the children see: 'Do all the pendulums swing in the same way?' 'What happens if you lift the piece of Plasticine up high and let go?' 'How high should you lift it up to get the best swings?' 'Whose pendulum swings for the longest?' 'Do some pendulums swing faster than others?'

Extension and differentiation

- Try counting in time with the pendulums to help children appreciate the speed at which they are moving.

- To help children understand that the length of string will influence the speed of the swings, use just one pendulum and ask them to clap in time with its movement. Repeat the process with different pendulums. Is the clapping faster, slower or the same?

- Extend the activity by asking the children to arrange the pendulums in order of length. Then ask them to arrange them in order of speed. What do they notice?

Experimenting with pendulums

Where in the room?

Group size:

up to eight children.

You will need:

a paperclip spinner marked with the following criteria: heavier, shorter, wider, narrower, longer, lighter (as shown).

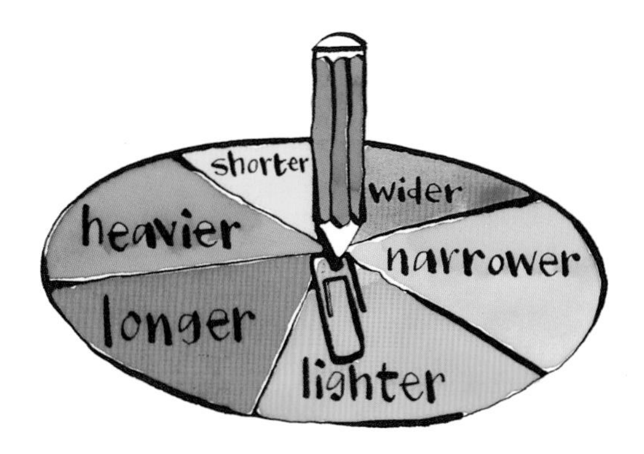

The activity

Start by discussing some of the things in the room in terms of their relative sizes ('Which things are longer than a paintbrush?' 'Which things are wider than the door?' 'Which things are narrower than the door?').

Invite a child to choose something in the room, then let them spin the spinner by flicking the paperclip round the pencil. Help them to read the word it lands on if necessary, then ask them to think of something that is wider/ longer/narrower etc. than their chosen item. Invite the other children to say whether they agree. Continue the game until all the children have had a turn.

Extension and differentiation

- To simplify the game, reduce the number of criteria on the spinner or use only one, such as weight.

- Introduce a scoring system by giving one point for each correct sentence.

- Let the children make a set of cards depicting things in the room and place them upside down in the middle of the group. Instead of choosing something in the room, ask the children to turn over the top card and spin the spinner. They must then think of something wider/longer/narrower etc. than the item shown on their card.

One cup

Group size:
up to ten children.

You will need:
transparent cups; plenty of other containers; water; a waterproof pen.

The activity
Fill a transparent cup with water. Then choose one of the other containers and ask children to guess how far up the second container the cupful of water will reach when it is poured in. Mark the children's estimates on the container with the pen, then pour the water in to check whose estimate was the closest.

Repeat the experiment with several different containers.

Extension and differentiation
- Support children by working with them individually or in pairs so that there are fewer marks on the container.

- Ask the children to estimate where three or four cupfuls of water will reach on the second container.

- Mark only one estimate that everyone in the group agrees with. The children will need to discuss, justify and negotiate their estimate.

See-saw

Group size:
up to eight children.

You will need:
a small see-saw or a 'home-made' one using a plank of wood over a log or similar.

The activity
Seat one average-sized child on one end of the see-saw. Let the other children take turns to sit on the other end of the see-saw and sort the group according to whether they are heavier or lighter than the first child (bearing in mind the sensitivities of any children who are significantly over or under the average weight for the age group). Discuss with the children how they know whether the children are heavier or lighter than the first child.

Repeat the experiment using another child to sit on the see-saw.

Extension and differentiation
- Reinforce the concept of 'lighter than' and 'heavier than' by allowing free play with balance scales.

- Ask the group to find out if anything changes when one person turns to face the other way. What about if one person puts their arms up high in the air? Does anything change if one person holds a bag of potatoes or a box of bricks?

- Ask the group to find out if taller children are heavier than shorter children. Are older children heavier than younger children? Will three children balance equally with one teacher?

● Present the children with similar things to compare on the see-saw, for example, the same amounts of wet and dry sand, or a box of folded newspaper and a box of loosely crumpled newspaper.

Evidence of the children's learning

There was a great deal of excitement during the *Pendulums* activity. The children really enjoyed using the pendulums, both with very long and very short strings. They freely used the appropriate language after encouragement. Three children were also able to explain that if the string was long the pendulum was swinging more slowly because it had further to go, which was a perceptive development!

The children found *One cup* particularly challenging, but certainly improved with practice. They volunteered opinions on each other's estimates, which occasionally led children to change their point of view. The widest use of language came when we tried to make a collective estimate that we all agreed on ('It will be full... No, it won't be enough to fill it... We might need to do it two times').

Assessment

● Are children able to use terms of comparison with accuracy?

● Are they able to locate objects that are longer, shorter, heavier, lighter, etc.?

● Are they able to sort items according to criteria of measurement?

Other quick ideas

● Use comparative language whenever possible ('Who takes longest to put on their coat?' 'Does it take longer to tidy up than it does to get things out?' 'Whose bag is heavier than Jane's?' 'How can we tell?').

● Stand a group of children in height order.

● Make up silly sentences together, for example: 'It takes longer to get to the moon than to eat my dinner!' 'An elephant is heavier than a mouse.' 'My toes are shorter than the paintbrush!'

Involving parents

Ask parents to set their children simple comparison tasks, such as finding out who is the tallest person at home, what are the heaviest and lightest foods in the cupboard (a bag of potatoes and one piece of pasta?), who can throw a ball the furthest, who takes the longest to get dressed in the morning, and so on.

How long? How tall?

Intended learning

To use the mathematical language associated with linear measure; to make direct comparisons of length; to find things which are longer or shorter than a given object; to order three or four objects by length.

Introduction

Length is usually the first type of measurement that children understand. It is the most straightforward because it is visual and involves working in just one dimension. However, children often find the language of linear measure confusing because there are so many words and contexts associated with it. (Length, height, distance, depth, width, circumference and perimeter are all linear measures.) When children come to use rulers, they encounter further pitfalls. Depending on the type of ruler, it is not always appropriate to begin measuring from the very end. And how should curved lines be measured? This chapter focuses on language associated with linear measure and the development of an appreciation that length is conserved even if the shape of a line is changed.

Key vocabulary

compare, long, longer, short, shorter, as long as, taller, tallest, shortest, longest, long enough

Plasticine snakes

Group size:

four to six children.

You will need:

Plasticine or clay.

The activity

Ask each child to make a 'snake' of Plasticine. Compare and discuss the lengths of the snakes, prompting the children with questions: 'How can we find out which snake is the longest?' 'How can we make the comparisons fair?' 'Who has a snake that is longer than mine?' 'Who has made the shortest snake?' 'Are there two snakes that are about the same length?'

Extension and differentiation

- To reinforce the concept, ask the children, in turn, to make up a 'longer than' or 'shorter than' sentence about two of the snakes, for example: 'Sarah's snake is shorter than Jamie's.'

- Ask the children to put the snakes in order of their length.

- Ask them to roll their Plasticine out until they think they have made a snake long enough to make a bracelet for themselves. Let them try the bracelet on to see if they were right. Try the same activity with lengths for a necklace, a ring and a belt. How different are the snakes in length?

- Ask the children to make snakes of the same length. Then ask them to curl and bend the snakes into different shapes, but without squashing or stretching them. Are they all still the same length?

Rolling out Plasticine to make snakes

Tall and short

Group size:
four to ten children.

You will need:
no special equipment.

The activity
Start by choosing two children of noticeably different heights. Let the rest of the group decide which of the two children is the taller, and which is the shorter. Emphasise the language used to describe the respective heights of the children ('Zoe is shorter than Ben. Ben is taller than Zoe'). Choose other pairs of children, and ask the rest of the group to make up similar sentences about them.

Extension and differentiation
- As a variation to reinforce the idea of comparing length, let the children find other things to compare and to make sentences about, for example: 'Charlotte's hair is longer than Samantha's.' 'Abdul's sleeves are shorter than Rohit's.'

- Invite three children to the front and ask the other children to make up sentences about them ('Sara is shorter than David, but David is shorter than Salheed. Salheed is the tallest').

Drip races

Group size:
up to six children.

You will need:
paint; one brush per child; an easel; paper.

The activity
Fasten the sheet of paper on to the easel. Let each of the children in the group choose a colour of paint and put some on a brush.

On the call of 'Ready, steady go!', let the children put a blob of their paint at the top of the paper and watch as it runs down. Discuss the resulting trails: 'Which line is the longest?' 'Whose line is longer than the blue one?' 'Who has made the shortest line?' 'Are there two lines that are about the same length?' 'Why do you think the lines are not all the same length?' etc.

Extension and differentiation
- To reinforce the concept of comparing length, put a sheet of paper on the table and let the children blow through straws on to blobs of paint to make the paint travel as far as possible. Compare the lines.

- Find a way to measure the trails. For example, cut pieces of string to the length of each trail and compare them.

- Experiment with different-sized drips, thick and thin paints, different starting places and so on.

Lengths

Group size:
up to six children.

You will need:
a ball of string; scissors.

The activity
Explain to the children that you need a length of string long enough to reach across the floor (or mat). Unravel the string so that it coils on the floor, and ask the children to guess when they think that the string is long enough. Cut the string and ask two of the children to stretch it out to check. Ask the group to say whether the string is long enough/too short/too long/ much too short/much too long/about half as long as it should be. Encourage as much use of language related to length as possible. Have several tries, and get the children to check each time by making comparisons with previous attempts until the string is the correct length.

Extension and differentiation
- To reinforce the concept of measuring, ask the children to work in pairs to make lengths of string long enough to measure round the table, the height of the teacher, etc.

- Use lengths of string to make an outline of someone in the group. Ask for a volunteer to lie on the floor with their legs and arms stretched wide, then allow another child to go round their outline with the string.

- Use lengths of string to make concentric circles, each one being slightly smaller or larger than the previous one.

Assessment

- Can the children tell you which of two (or more) paint trails or lengths of string is the longest and which is the shortest?
- Do they line up the items to be measured?
- Given a choice of objects, can they find those which are longer or shorter than the one you have chosen?
- Do they use the language (longer, shorter, taller, longest, tallest, shortest) appropriately?
- Can they order several objects from shortest to longest, or shortest to tallest?

Evidence of the children's learning

The children loved *Drip races* and we played several times. It was a lovely context for children to use the language of length (long, longer, etc.) and the language of distance (further, not as far, etc.) alongside each other.

They found the *Lengths* activity particularly challenging, and often misjudged the lengths. However, they went on to use the incorrect lengths to help them make the next estimate. Right or wrong, the children certainly used the language of length a great deal, correctly in this context.

Other quick ideas

- Ask the children to hunt round the room for three things that are longer or shorter than their shoe.
- Get the children to stand in a row in order of height. Think of one of them and let the children ask questions to help them identify who the person is: 'Are they taller than Jan?' 'Are they shorter than Tom?'
- Let the children cut five different lengths of string and put them in order.
- Ask the children to use a ruler to draw straight lines of different lengths.
- Challenge them to thread beads to make a string the same length as your arm.

Involving parents

Ask parents to help their children cut pieces of string to the same length as the height of each member of the family. Suggest that children could be challenged to find things that are shorter and things that are longer than, for example, an old tie. Allow free play with a tape measure. Discuss what the child finds out.

Fill it up

Intended learning

To fill and empty containers; to describe containers as full, half full or empty; to begin to develop ideas of displacement; to guess and check how full a container will be after a given amount of liquid is poured into it; to describe a container as holding more or less liquid than another container.

Introduction

Capacity is a difficult concept for young children to appreciate. Most children will choose the tall, thin glass of drink in preference to the short, fat one, even though the short, fat one may hold more! They can also think that a quantity of liquid changes if it is poured from a short, fat glass to a tall, thin one. Another difficulty they find is in distinguishing between the space something takes up (its volume) and what it will hold (its capacity). Displacement is used to measure volume and refers to how much the water level rises when a solid object is placed in a container. Children's most common experience of this is how the level of the bathwater rises when they get in.

The following activities give children experience of capacity and water displacement in a structured play environment.

Key vocabulary

full, empty, container, half full, fuller, pour, overflow, liquid, flow, fast, slow

Trickle lines

Group size:
up to six children.

You will need:
a plastic bottle with a small nozzle (such as a washing-up liquid bottle) for each child; a large bucket of water; a concrete playground or similar outdoor surface.

The activity
Ask the children to fill the plastic bottles with water and then let them trickle the water out to make a long trail on the ground. Discuss the trails: 'Whose is longest?' 'Why do you think that might be?' 'Who has made a straight trail?' 'A curved one?' 'One with corners?' 'Is it always the largest bottle that makes the longest line?' 'Does the size of the nozzle make a difference?'

Extension and differentiation

- Support the children by helping them to compare their lines (using a tape measure or a length of string, for example).
- Let all the children try to make the longest line they can.
- Direct the kind of lines the children try to make, for example, a straight line, a curved line, the first letter of their name, a shape such as a triangle, or a number.

The children enjoyed dropping stones into a pot of water.

Stones

Group size:
up to eight children.

You will need:
plastic containers in a range of sizes (transparent ones are best); water; some pebbles.

The activity
Fill a pot nearly to the top with water. Discuss with the children what they think will happen if one of the stones is placed in the water. (They may suggest that it will sink, that it will splash, or that the water will flow over the top.)

Place a stone gently in the water. Do they notice anything different? Discuss what might happen if more stones are put in the water. Keep adding stones until the water flows over the top of the container.

Extension and differentiation
- Let the children experiment for themselves as necessary to reinforce the activity.
- Try the activity again, but ask the children to guess in advance how many stones might be needed to make the water overflow.
- Reverse the activity. Nearly fill a vessel with the stones, and then predict how much water will be needed to make it overflow.

Holes

Group size:
four or five children.

You will need:
up to ten plastic bottles; a skewer; water.

The activity
Before the activity, use a skewer to make some small holes in the bottles at different levels. Put different numbers of holes (between one and five) in each.

Ask the children each to choose a bottle and fill it with water, and then to watch closely as the water flows out of the holes. Discuss what happens to the water. Does it always come out in the same way? *(It actually slows down and the arc of the flow changes as the bottle empties.)* Why don't the bottles empty? *(The water stops draining at the level of the lowest hole.)* What happens when the children cover the top of their bottle with a hand as the water is trying to escape through the hole? *(The flow slows down or stops.)* Do the bottles with most holes empty most quickly? What happens when the children blow into the top of the bottle? *(The water flows more quickly.)* Does it make any difference if they tip the bottle slightly?

Extension and differentiation
- Allow plenty of free play with the bottles to familiarise the children with what happens.

- Let each child in the group choose one of the holed bottles and fill it with water. Ask them to hold their hands over the top when their bottle is full, then on the cry of 'Ready, steady, go!', take their hands away and let the bottles drain. The first bottle to drain to the lowest hole wins the race.

- Let each child choose a different bottle and fill it. Ask the children to try to get all the bottles to finish draining to the lowest hole at the same time by using their hands over the top to control the flow. (This stops the flow because air can no longer get in to fill the space that the water would leave.)

Overflow

Group size:
up to six children.

You will need:
a selection of plastic bottles of various shapes and sizes; water; a funnel (optional).

The activity
Begin by demonstrating how much water the biggest, tallest and fattest bottles can contain in relation to the smallest, shortest and thinnest. Show the children how to pour the water from one container to another without spilling any (a funnel would be useful).

Ask the children each to choose a bottle and fill it with water. Now ask them to choose another empty bottle that they think would overflow if they filled it with the water from their first bottle. Were they right?

Extension and differentiation
- Try the same activity with different types of containers, such as yoghurt pots.

- Use rice, sand or gravel instead of water.

- Ask the children if they can find two bottles that hold about the same amount of liquid. Let them check by pouring from one to the other without spilling any.

Assessment

● Can the children tell you when bottles filled to various levels are full, empty and half full?

● Do they recognise that spilling changes the outcome when trying to find out if one bottle holds more or less than another? (This can be tested in a fun way by saying that your small bottle will hold more than their large one, then pouring from the large to the small, but spilling lots of water on the way. The reactions will indicate the level of their understanding.)

● Can they explain that putting something in a container of liquid will change its level?

● Are they beginning to appreciate that two bottles might be different shapes, but can still hold similar amounts of water? Or that a tall, thin container does not necessarily hold more than a short, fat one?

Evidence of the children's learning

The *Holes* activity delighted the children, and they began to freely use language associated with capacity, particularly when they were asked to find out which bottles drained for the longest. They were anxious to fill them right to the top, and discussed how difficult it is to 'get them right full without spilling it'. There were also many reports that the same activity was being tried at home at bathtime!

Other quick ideas

● Fasten elastic bands at different levels around several bottles. Ask the children to fill the bottles with water up to the elastic bands.

● Colour water with paints of various colours. Let the children make 'magic potions' with different amounts of each colour, for example, three spoonfuls of blue paint, five of yellow paint and one of red paint.

● Choose a small and a large bottle. Fill the small bottle with water and ask the children to predict how far up the side of the large bottle the water will come when poured into it.

● Challenge the children to fill three pots to the same level.

● Ask the group to fill some containers as full as possible without them overflowing.

Involving parents

Ask parents to make available a variety of large and small bottles to play with in the bath, and let their children practise pouring generally. They could also make a funnel from the top half of a plastic bottle. Suggest that they provide yoghurt pots to act as cups, and challenge their children to pour the same amount of water for each of the teddies, dolls or other toys. Parents could also be encouraged to discuss with their children how many bottles or cartons of drink will be needed for a party, or for the week.

How heavy?

Intended learning

To use and understand the mathematical language associated with weight; to make a direct comparison between the weight of two objects; to find things heavier or lighter than an object; to make estimates of comparative weights by holding one object in each hand.

Introduction

There are two words associated with measuring how heavy something is: mass and weight. Mass is a measure of how much matter is contained in something. When gravity is exerted on mass, it has weight. There is much debate about which term children should learn. In everyday language it is more convenient to talk about weight, because we are unlikely to measure mass in a gravity-free environment.

The activities in this chapter help children to appreciate the role of a balance, and what it means if one side goes down when objects are placed in the pans. However, a balance has the limitation of only being able to compare the weights of two objects. These activities also introduce the much more sophisticated task of ordering three or more objects using a balance. Before attempting any of them, allow the children an opportunity to play freely with a balance and a range of items to weigh.

Key vocabulary

compare, heavy, heavier than, light, lighter than, as heavy as, lightest, heaviest, balance

Is it heavy or is it light?

Group size:

up to ten children.

You will need:

a box containing an interesting collection of objects, such as toys, crayons and bricks; a table labelled 'heavy' at one end and 'light' at the other; a pan balance.

The activity

Tell the children that they are going to sort the objects according to whether they are light or heavy and that they must put them on the correct side of the table. Let each child choose something from the box to put in the right place on the table. Explain that any time they want to use the balance to check, they can.

When they have distributed all the objects, invite a child to find someone with an object 'heavier' than theirs. Let them check using the balance. Say to the child, 'So your object is light.' Then ask the same child to find someone with an object 'lighter' than theirs. Again let them check on the balance. Say to the child, 'So your object is heavy.' Invite the group to suggest which is right – is it heavy or is it light? Discuss how things can be heavy or light, depending on what they are compared with.

Extension and differentiation

- If children are struggling to understand the concept, help them to place three objects in order of weight, then make up sentences about them ('The crayon is lighter than the brick, but the brick is lighter than the doll').

- Put one object (such as a book) in the centre of the table and change the labels to 'heavier than the book' and 'lighter than the book'. Now ask the children to sort the objects again.

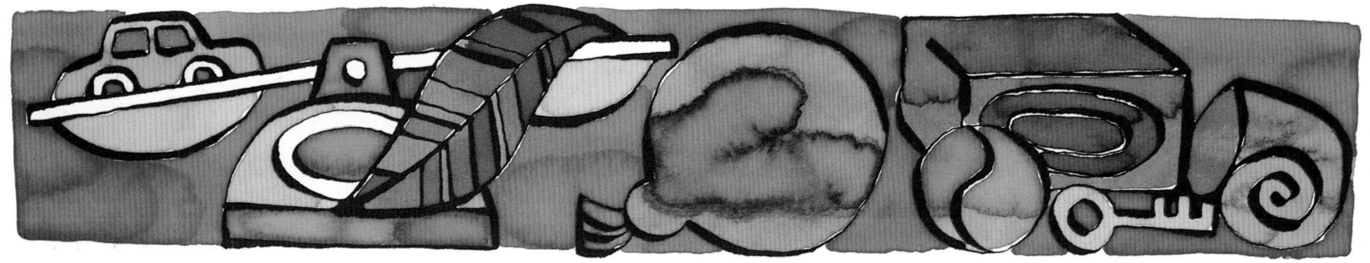

Heavy teddy

Group size:
four to six children.

You will need:
a balance; a small teddy bear; two trays labelled 'heavier' and 'not heavier'; six or seven other objects for weighing, such as a ball of wool, a key, a conker, a feather, a shell, a stone, a brick, a ball or an empty box.

The activity
Make sure the children appreciate that if one object is placed in each pan of the balance, the pan that drops contains the heavier object.

Place the teddy in one of the pans. Ask the children to balance each of the objects against the teddy, then place them in the appropriate 'heavier' or 'not heavier' tray.

Extension and differentiation
- If the children are not sure how the pan balance works, give them an opportunity to balance pairs of objects and tell you which is heavier and which is lighter.

- Give the children a chance to hold the teddy and the other object before putting them in the pans. Ask them if they think it is going to belong in the 'heavier' or the 'not heavier' tray.

- Replace the teddy with one of the other objects provided.

The children weighed an assortment of objects against the teddy bear.

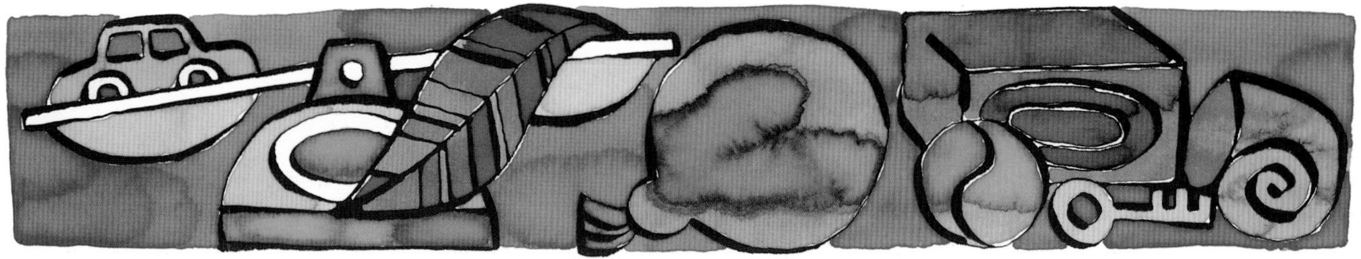

Stones

Group size:
up to six children.

You will need:
a balance; a varied collection of smooth stones (alternatively, substitute the stones with small model toys or bricks).

The activity
Let the children take turns to pick two stones, and try to guess what will happen when they place one in each side of the pan balance. Discuss the possibilities with them: 'How will we know which is heavier?' 'If one side is heavier, what is the other side?' 'What will it mean if the balance doesn't tip either side?'

Let the children try putting their stones in the balance pans. What happens? Were their predictions correct?

Extension and differentiation
- To make the activity easier, give the children two stones of obvious contrasting weight.

- Try putting two or three stones in each side.

- Show the children one big stone. Ask them how many small stones they think they would need to put in the other side of the balance to make the two pans level. Place the big stone in one side of the balance, and ask a child to put small stones one by one in the other side until the pans balance. It is likely that the children will want to continue until the pan with the small stones is the heaviest.

Balloons

Group size:
up to ten children.

You will need:
a pan balance; some balloons.

The activity
Blow up one balloon. Ask the children, 'If I blow the next balloon up so that it is bigger than this one, do you think it will be heavier or lighter?' Inflate the second balloon so that it is bigger and put the two balloons in the balance pans. Discuss the outcomes. Which balloon do the children think was heavier? Why do they think that is?

Let the children try comparing a range of balloons inflated to different sizes. Include some that haven't been inflated at all.

Extension and differentiation
- Support children by providing them with balloons inflated to different sizes. Let them watch you inflate them so that they can see that the balloons are essentially the same, even though the amount of air inside is different.

- Try filling a balloon with water. (This is probably best attempted in the playground.) Ask the children if it will be heavier or lighter than one that has been blown up. Does the fact that it is smaller and more wobbly mean that it will be lighter? Offer the children a selection of balloons of all sizes, some filled with water and some with air, and give them the opportunity to experiment for themselves.

Assessment

- Can the children tell you what a pan balance is for?
- Can they tell you which of two objects is heavier and which is lighter, using the pan balance to compare the weights?
- Given one object in one side of a balance, can they find another which is heavier, and one which is lighter?
- Can they use the language associated with weight in a context, for example, can they tell you that one object is heavier than another?

Evidence of the children's learning

The children had always enjoyed playing with the balance. However, it was obvious that after they had experienced these activities, their play changed. When they were playing freely, there was little discussion other than 'Watch this!' when something really heavy was placed in one pan, to make it crash down. After these activities, the children still enjoyed the crashing down of a pan, but would talk about objects being 'really really heavy' or 'heavier than all the others'.

Other quick ideas

- Make available a balance and a range of interesting objects. Encourage the children to balance single objects against each other (such as a toy car against a teddy), and single objects against multiple ones (such as a toy car and some interlocking cubes).
- Practise guessing. Ask the children to hold two objects, one in each hand, and guess which is lighter, then check using the balance.
- Let the children try to balance a ruler across a finger. Lead them to understand that the finger needs to be in the middle of the ruler.

Involving parents

Suggest that parents play on a see-saw with their children and discuss how it is possible to go on it with someone heavier, provided they sit nearer the middle. At home they could use old pieces of wood or card to make a see-saw for some toys. Parents could also be asked to play with a balance with their children and discuss what they find out.

When?

Intended learning

To use and understand the language associated with time; to begin to understand the cyclical nature of time; to identify events which occur before or after others.

Introduction

Time is difficult for children to understand because it is completely abstract, unlike measures such as length, where children can actually see what they are measuring. Also, confusion results from the misuse of language in everyday speech; we say things like 'Hang on a minute' and 'Just a second', when we just mean 'Wait!' There are also many terms that can mean different things in different contexts; 'soon' in 'Christmas will be here soon' implies a different time span from 'It's tea-time soon!'

There are two aspects of time that children need to learn. Firstly, they need to understand the passage of time and associated language. This incorporates sequencing events. They also need to learn to tell the time by using clocks and, again, the associated language. This chapter focuses on the first of these aspects.

Key vocabulary

before, after, next, at the same time, today, tomorrow, yesterday, days of the week, how old, birth-day, soon, minute, day, second, early, earlier, late, later

Before and after

Group size:

up to eight children.

You will need:

no special equipment.

The activity

Ask the children to sit in a circle. Invite them to think of things that they do or things that happen over the course of the day (cleaning teeth, the postman comes, having a bath, etc.). Ask one child to think of one such event, either one that has been discussed or something new. Then ask the child sitting to their right to think of something they would do before this, and the child to the left to think of something they would do after this. For example, the first child might suggest 'watching television', in which case the child on the right might then say 'get home from school' and the child on the left might say 'eat my tea'.

This activity often leads to discussion of what is done before and after daily activities, for example, whether the children wash their face before or after cleaning their teeth, and so on. This might vary from household to household.

Extension and differentiation

- To reinforce the activity, ask the children to draw pictures of 'Things we do in the morning' and use them for a display. Do the same for afternoon and evening events.

- Ask the children to draw pictures of three things (or more if appropriate) they do during the day, and put them in order.

Pegs

Group size:

up to 12 children.

You will need:

drawing equipment; a length of string tied tight across the room between the backs of two chairs; clothes pegs.

The activity

Tell the children a familiar story, such as Goldilocks. Discuss the sequence of the story and invite individual children to draw pictures of the main events (Goldilocks walks through the forest, she finds a cottage, and so on).

Ask the children, one by one, to peg their pictures to the length of string in the order that the events happen in the story. This may mean that pictures need to be moved along as new ones are pegged up. Emphasise the language of time – before, after, at the same time, next.

Go through the story with them again, asking them to point at the correct pictures as you go along to check that they are in the right order.

Extension and differentiation

- Reinforce the concept of sequencing by repeating the same activity but using events in the children's day (or week) instead of a story.

- Have fun mixing up the pictures and telling the story in the wrong order.

Sequencing events from the story of Goldilocks

Birthdays in order

Group size:
four to six children.

You will need:
writing and drawing equipment; sticking putty; display materials.

The activity
Begin the activity by discussing birthdays. Does anyone know when their birthday is? Which month is it in? Which date?

Ask each child to draw a self-portrait, then help them to write the date of their birthday underneath. Collect together the pictures of any children whose birthdays are in the current month, and display them on a chart along one wall, starting with the nearest birthday to the present date. Use sticking putty so that the pictures can be easily removed. Continue with all the other months in order.

During the year, as each birthday passes, move the pictures from the beginning to the end of the line, so that the children always know whose birthday is next. Refer to the chart every day, counting down to the next birthday.

Extension and differentiation
● To reinforce the activity, let the children make a small version of the chart, showing the birthdays of the people in their family.

● Make a similar chart showing events that will happen during the day (drawing, music, playtime, and so on). Invite individual children to move the pictures to the back of the line as the day goes by.

In a minute

Group size:
up to ten children.

You will need:
a variety of equipment for timing (a watch, a clock, a sand timer, a cooking timer, etc.).

The activity
Discuss with the children what is meant by 'a minute'. Explain that they are going to guess how long a minute is.

Let them watch each of the timing devices run for one minute. Then hide the timers from the children, but place one so that you can see it, and set it going for one minute. Ask the children to indicate (put their hand up, stand up, clap or put their hands on their head) when they think one minute has passed. Play this a few times with different timers. When using a watch or other silent timer, you will need to say 'now' to let the children know when the minute is through.

Extension and differentiation
● Some children may have difficulty in focusing on the time passing for a whole minute. In this case, reduce the time to, say, ten seconds. The children can be encouraged to count the passing seconds.

● Set the children challenges ('How many times can you smile in a minute?' 'How many times can you write your name in a minute?' 'How many balls can you place in a bucket in a minute?').

Assessment

- Do the children know what a clock, timer or watch is for?
- Can they put two or three events in order of when they happen?
- Given one event, can they think of something that happens before and something that happens after?

Evidence of the children's learning

There is always great excitement when birthdays are discussed and the *Birthdays in order* activity was no exception. The chart helped the children sequence these events, and they used it of their own accord to discuss whose birthday (or more specifically whose party!) was next, and who had the longest time to wait.

In *Pegs*, the visual sequencing of the story was useful in developing the skill of recounting a story, and the children enjoyed using the 'peg pictures' as a prompt.

Other quick ideas

- Involve the children in daily planning, and set aside a time for discussing what you will do each day ('We'll do some drawing, then build with bricks, then we'll play and then have a story'). This could also be recorded in pictorial form for the children to refer to.
- During the course of other activities, groups of three or four children could take turns to use timers set at five-minute intervals to observe and tell you what happens during each five minutes. Building and making activities are perhaps the easiest to observe in this respect.
- Set a timer to ring or buzz every hour throughout the day. Recap with the children on what they have done in that hour.

Involving parents

Ask parents to involve their children in making plans for the day at home, and to talk about what the clock will look like when it is time for tea, Blue Peter, going out, going to bed, etc. Parents could also draw pictures of clocks showing these various times and place them next to a real clock for comparison. Encourage them to chant the days of the week and the months of the year with their children.

Boxes

Intended learning

Junk models: to find out about conservation of measure and shapes at different orientations.
Tall towers: to use the language of length.
Covering boxes: to recognise the faces of three-dimensional shapes.
Fit it in: to use the language of position.

Introduction

This chapter, and the three following chapters, take a thematic approach to space, shape and measure.

Empty boxes are a wonderful resource for this area of mathematics. Children can cover them, and learn about length and area; they can fill them, and learn about capacity; they can build with them, and learn about length, area, volume and shape; or they can print with them, and learn about shape. Before starting these activities, allow the children to play freely and to build models with the boxes.

Key vocabulary

large, small, turn, face, upside down, facing, opposite, circle, triangle, rectangle, cube, cuboid, cylinder, prism, tall, taller, tallest, short, shorter, shortest, inside, wide, wider, narrow, narrower

Junk models

Group size:

two or more children.

You will need:

identical sets of assorted boxes for each child (for example, a toothpaste box, an egg box and a cereal box); sticky tape; glue; paints and paintbrushes.

The activity

Give each group a similar set of boxes and ask the children to make a model. Encourage them to use all their boxes.

Help the children to join the boxes together with sticky tape or glue as appropriate, then allow the children to decorate their models.

When any paint has dried, discuss with the children what each of the boxes has been used for, what it is joined to and which way round it is.

Extension and differentiation

- Allow the groups to choose two sets of identical boxes and to make two different models.
- Let the children exchange models and try to identify the boxes hidden under the paint.

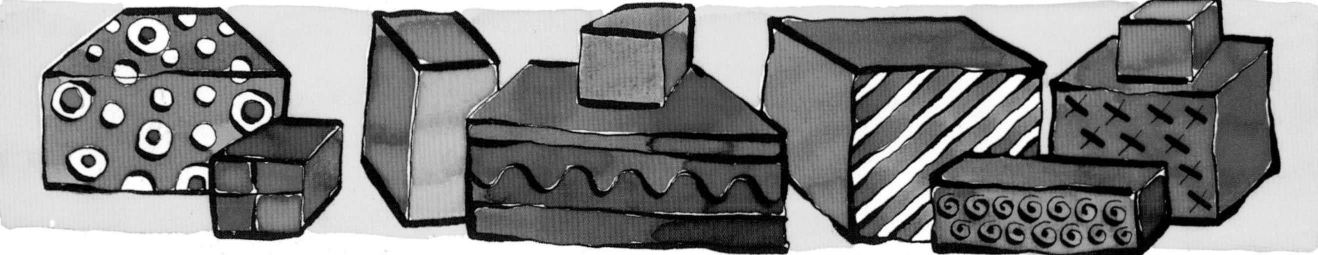

Tall towers

Group size:

up to six children.

You will need:

a varied selection of boxes; glue; sticky tape; painting and drawing equipment.

The activity

Explain to the children that they are going to build towers. Let each child choose six boxes, then ask them to fasten the boxes together to make the tallest tower that they can. Compare the resulting towers by asking the children questions such as: 'Which is the tallest?' 'Which is the shortest?' 'Are there two the same height?' 'Could any have been taller?'

Ask the children to put the towers in height order. Display them with appropriate captions: 'Sangita's tower is bigger than Stuart's', 'Kelly's tower is the shortest,' etc.

Extension and differentiation

- Ask the children to choose a different set of boxes and use them to make the shortest tower they can, or perhaps a tower that matches their own height.

- Allow each child to choose one more box and add it to their tower. Is the tallest tower still the same one?

The children were allowed to choose boxes for building tall towers.

Covering boxes

Group size:
four to six children.

You will need:
boxes; old gift wrapping paper; scissors; glue.

The activity
Explain to the children that they are going to choose a box and cover it with paper. Make sure they understand that this differs from wrapping parcels with just one sheet of paper; here they are going to cover each separate face of their box with paper cut to fit.

Encourage them to work out their own strategies. They might, for example, cut a shape and then see if it fits; tear off small pieces and stick them on one at a time; cut a big shape and then cut the extra off once it has been stuck in place; or draw round each face on to the paper and then cut it out. As they work and when they have finished, ask them to explain their method to the other children. How many pieces of paper did they need? Were any of the pieces of paper the same size? What shapes were the faces on their box?

The decorated boxes may then be used for other activities, such as building a maze (see page 25).

Extension and differentiation
● Encourage the children to draw round and cut out a shape for each face. Discuss the shapes they have created before they stick them on to the box. Does anyone else have a similar set of shapes?

● Let them open the boxes out flat. Do they still need to use the same amount of paper to cover the boxes?

● Attach the paper with removable putty, then remove it to see what it looks like flat.

● Give the children unusual shaped boxes to cover (circular cheese triangle boxes, unusual shaped chocolate boxes, etc.).

Fit it in

Group size:
up to six children.

You will need:
a wide selection of boxes in a range of sizes.

The activity
Ask each child to choose a big box, and to take turns to choose another box that will fit inside the first one, then one to fit inside the smaller box, and so on. Discuss the task as they work ('Why did you choose that box?' 'Could you find a bigger one that will fit in?' 'What is the best shape to choose?' 'How many boxes have you used?').

Extension and differentiation
● Reduce the number of boxes available if necessary to make the activity simpler.

● Ask the children to remove the fitted boxes from inside each other, and then make the tallest tower they can. What is the best order for assembling the boxes?

● Make the activity competitive by asking the children to fit as many boxes together as they possibly can.

● Extend the activity so that the boxes do not need to fit directly inside each other, but can be arranged in layers, two or three at a time.

Assessment

Specific skill assessment ideas are covered fully in other units and can be applied to these activities, as appropriate. However, watching and listening to children as they take part in these activities also offers the opportunity to:

- identify whether or not skills developed in activities from earlier units have been remembered and are being used;

- identify whether or not the children need further experience of activities in other units to extend their understanding. For example, if a child has difficulty with the *Tall towers* activity it may be useful for them to try some of the activities in the chapter, *How long? How tall?* (pages 32–35).

Evidence of the children's learning

The children were genuinely surprised at the variety of models that could be made with the same set of boxes. They engaged in unprompted discussion about the different orientations of the boxes, and the different uses they had been put to. They even commented that some of them 'look different sizes' as part of a model.

Covering boxes was a great stimulus for talking about rectangles and squares, and differences between them. The children talked about rectangles being 'long squares'. One girl identified the fact that there are 'all different sorts of rectangles, but squares are all the same really, just bigger or smaller'.

Other quick ideas

- Ask the children each to describe the outside of their home. (This discussion may take a few days to complete as it is important that the children all have a good look!) Let them use boxes to make a model of their home.

- Choose three boxes. How many different rectangles are there? Draw round them to check.

- Wrap up a box as if it were a present.

- Fill boxes of different sizes with balls of the same size. Count the balls in each one.

- Cut a box into individual faces. Name the shapes. Are there more rectangles or squares?

Involving parents

At birthdays, Christmas and other festivals, parents can involve children in wrapping parcels by choosing paper the correct size and learning how to fold it. Other activities might include filling empty boxes with different things (sand, stones, screwed up newspaper, etc.), making junk models with boxes, or making holes in smaller boxes and threading them on string to make snakes.

Balls

Intended learning

Paint balls: to make a connection between circles and spheres, and to recognise that circles and spheres are the same, whatever the orientation.
Tell me about it: to use the language of shape and measure, and to explore how to classify objects by their properties.
Marbles: to use the language of shape.
Rolling: to use the language of length.

Introduction

Balls hold a particular fascination for children, so it is worthwhile to tap that enthusiasm and focus their play to support the development of some mathematical ideas.

Spheres are unique three-dimensional shapes in that they have only one face. Every point on that face is equidistant from the centre. The following activities give children experience of the special nature of these shapes.

Key vocabulary

turn, face, rectangle, sphere, circle, triangle, curved, straight, upside down, next to, long, longer, longest, cross, further, furthest

Paint balls

Group size:

up to six children.

You will need:

a selection of plastic balls; sponges; paint; large sheets of paper.

The activity

Soak the sponges in paint of various colours. Show the children how to press the balls on to the sponges, and then on to paper to make a print. Then let them work on their own or in pairs to use this technique to make a picture.

Ask the children questions about the pictures they make with the balls: 'What shapes do the balls make?' 'How do you know it's a circle?' 'Are there any that are not circles?' 'How many have you made?' 'Can you do another picture so that each circle is in a space of its own?' 'Can you do another picture so that all of the circles overlap?' 'Which is the biggest circle?' 'The next biggest?' 'The smallest?'

Extension and differentiation

- Allow free play with the paint and balls to help develop children's understanding.

- Give the children long sheets of paper and ask them to print a line of circles that get progressively larger or smaller.

- Can they print different-sized circles with the same ball?

- When the paint has dried, draw a different face in each circle.

Tell me about it

Group size:

up to eight children.

You will need:

a range of balls of different sizes, patterns, textures, etc.

The activity

Introduce the activity by showing the children one of the balls. Explain that they are going to think hard about the ball and to find out as much as they can about it. Ask them a few questions, such as: 'What colour is it?' 'What do you think it is made of?' 'How far do you think it would roll?'

Invite the children to ask questions or volunteer some information about the ball. They may choose to talk about size, colour, texture, hardness or softness, shape, how high it will bounce, how it can be used, whether or not it would be easy to catch, etc.

Finally, let the children sort the balls according to one of these criteria.

Extension and differentiation

- Children can be supported by developing the activity into a game where a ball is passed around the group, and each child must say something new about it when they hold it.

- Take the whole group outside with a large selection of balls and try a range of experiments. How many balls can each child hold? How far can they roll a big ball? A small ball? Can they blow any of the balls along? Can they kick a ball further than they can throw it?

Sorting balls according to size

Marbles

Group size:
four to six children.

You will need:
a selection of marbles; paint in shallow containers; paper; trays with reasonably high edges (one per child).

The activity
Demonstrate how to place a sheet of paper flat in the bottom of a tray, dip a marble into some paint and place it in the tray, tipping the container carefully to make a trail of paint on the paper.

Give the children a lined tray each and let them experiment using different-sized marbles and different colours of paint. As the children work, discuss what they are doing: 'Can you make straight lines?' 'Can you make curved lines?' 'Can you make a line from one side of the paper to the other?' 'Can you make a line that goes all the way around the edge of the tray?' 'Can you make a line that goes from one corner to the other?' 'Which marble made the longest line?' 'Which made the shortest?' 'Tell me about your other lines,' 'Why do you get lines when you have used a ball shape?'

Extension and differentiation
● A variation to reinforce the activity would be to use larger balls dipped in paint and roll them around on huge sheets of paper placed on the ground outside.

● Set the children challenges. Can they make a picture that has only straight lines? Only curved lines? Only six paths across the paper?

Rolling

Group size:
up to six children.

You will need:
a collection of balls, all different sizes (and different shapes, such as rugby balls, if available); a slope (for example, a small slide, or a propped-up plank of wood or board).

The activity
Let each child choose a ball and roll it gently down the slope. As they do so, discuss how far the balls roll, using the appropriate language, for example, further, faster, furthest, fastest. When all the balls have been rolled, ask the children each to go and stand next to theirs. Ask them to identify which one is first, second, third and so on, according to how far from, or close to, the bottom of the slope they are.

Extension and differentiation
● Reinforce the activity by repeating it, with the children choosing different balls.

● Ask the children to arrange the balls in order of how far they rolled. Do the bigger balls roll the furthest?

● Change the steepness of the slope. Do the results stay the same?

Assessment

Skills assessment for these activities is covered in the chapter, *Circles* (page 11). However, watching and listening to the children also offers an opportunity to:

● identify whether or not the children have made a connection between spheres and circles, and whether or not they are beginning to understand the properties of these shapes;

● identify whether or not skills addressed in other units have been remembered and are being used, and whether or not the children need experience of further activities to extend their understanding. For example, if a child has difficulty with the *Rolling* activity, it may be useful to try some of the activities in the chapter, *How long? How tall?* (pages 32–35).

Evidence of the children's learning

One of the children who was involved in the *Paint balls* activity gave a wonderful description of the connection between circles and spheres. As he worked, he declared that 'Balls are just puffed up circles really, aren't they?' This led to a discussion about the word sphere as the proper word to describe the shape of the balls, and how it could also be used to describe the shape of the earth.

The children also talked about things which were 'nearly spheres', such as apples and conkers. When they were asked why they weren't really spheres, they identified the fact that conkers were often a bit 'knobbly' and that apples have 'that hole in them' where they were joined on to the tree!

Other quick ideas

● Let the children play with balls in the sandpit. Encourage them to press the balls into the surface of the sand and lift them out carefully to make a 'lunar landscape'.

● Find out which ball bounces the highest.

● Provide an assortment of circular containers (yoghurt pots, ice cream tubs, margarine tubs, and so on). Ask the children to find balls that will fit snugly into each one.

● Let the children play an 'over and under' game in two teams. Ask them to stand in two lines, one behind another, about half a metre apart. Give a medium-sized to large ball to the first child in each team, and get them to pass it over their head to the next child, who then passes it between their legs to the third child, and so on to the end of the line. The last child races with the ball to stand at the front of their line. The first team to get their ball back to the front of their line 'wins'.

Involving parents

Parents could be invited to look at bubbles with their children and talk about the shape. Children could be allowed to observe what happens when balls of biscuit dough are cooked. Other activities could include making different-sized balls of Plasticine, or finding different-sized balls around the house (fluffy bobbles on hats and slippers, sweets, etc.).

Line up

Group size:
up to ten children.

You will need:
some chalk or small pieces of masking tape.

The activity
Tell the children that they are going to line up, but before they do, you would like them to guess how long the line will be. If the line begins at the door, where do they think the last person will be standing? Let each child or pair of children use chalk or tape to mark where they think the end of the line will be, asking questions as appropriate ('Would we reach my desk?' 'Would we reach the opposite wall?').

Next ask the children to line up to find out whose guess was closest. Encourage them each to say a sentence to describe their guess, for example: 'I guessed that the line would be a little bit longer.' 'I guessed that the line would be a lot shorter.'

Extension and differentiation
- Reinforce the activity by asking the children to line up in a different order. Is the line the same length?

- Try other ways of getting in a line: side by side shoulder-to-shoulder; lying down end to end; next to each other with both arms out-stretched, finger tip to finger tip. As before, ask them where they think the line will reach in the room, then check their guesses. Compare these lines with the original one. Are they longer or shorter?

- Ask the children to find a way to record the outcomes to make a display.

Car tracks

Group size:
up to six children.

You will need:
old toy cars; shallow trays of paint; large sheets of paper.

The activity
Show the children how to 'drive' the toy cars through the tray of paint, then across a sheet of paper. Then let them experiment with their own sheets of paper and vehicles. (The cars will probably need to be washed between colours.)

When the paint is dry, discuss the resulting lines on the paper ('What do you notice about the lines the car tyres have made?' 'Do any of the lines move closer together or further apart?' 'Can you see where the car tracks crossed?' 'Did any cars go round a corner?' 'Can you tell how fast the cars were going?').

Extension and differentiation
Set the children differentiated 'car driving' challenges according to their abilities:

- Drive the car so the tracks make a curved, straight or wavy track.

- Drive the car to make straight tracks across the whole page.

- Drive the car to make a triangle.

- Make tracks that do not cross any others.

- Make tracks that all cross at least one other.

Assessment

Assessment of the skills addressed in this chapter is covered more fully in *Introducing shape* (page 7) and *How long? How tall?* (page 35). However, watching and listening to children as they take part in these activities is a good environment for identifying:

● whether they make a connection between shapes and the number of sides they have;

● their use of the language of position and the language of movement;

● whether skills addressed in other units have been remembered and used;

● whether the children need some experience of activities in other units to extend their understanding. For example, if a child has difficulty finding shapes in *Stringy lines*, it may be useful to try some of the activities in the chapters *Introducing shape, Circles, Triangles* and *Squares* (pages 4–19).

Evidence of the children's learning

The discussion in *Car tracks* and *Stringy lines* was rich with the language of measure, shape, direction and movement, both as the children worked and afterwards. They enjoyed looking at each other's pictures and trying to describe the routes that the cars or the string had followed. They were keen to spot similarities with and differences from their own ('I did a long green line too!' 'My yellow wool is wiggly!'). There was also great excitement when children noticed triangles in the pictures; they were very particular about the fact that the sides of the triangles had to be straight!

Other quick ideas

● Let the children practise drawing straight lines with rulers.

● Ask them to draw ten lines anywhere on the paper, with some of them crossing each other. What shapes can they see among the lines?

● Draw lines in the air in front of the children and ask them to copy or mirror what you do. Start with straight lines in several directions, to either side and up and down. Try some that move towards you. Start by pointing at the children and then move your finger so that it points at you. Do the children copy you? Try curved lines in the air and two crossed lines.

● Let the children take turns to come out to the front and draw different kinds of lines (straight, curved, circular, zigzag, wavy, spiral, etc.) on a large sheet of paper.

Involving parents

Invite parents to challenge their children to find as many different lines around the home as possible. These might be washing lines, telegraph lines, lines in the tablecloth, and so on. Suggest that they help their children to make some long biscuits, some wavy, some curved, some straight, some with corners, etc. Children could also be allowed to line up objects at home, such as cutlery, fruit or socks and then be asked questions about them, such as: 'Which makes the longer line – the spoons or the forks?'

Bags

Intended learning

Make your own bag: to use the language of measure; to recognise and name shapes; to observe symmetries.
Sort: to use the language of shape and measure.
Cotton wool balls: to measure capacities. *Dressing-up bags:* to compare sizes.

Introduction

Apart from the creative uses of paper bags (puppets, masks, hats), paper bags have a wealth of potential for developing mathematical ideas. Children can fill them, and learn about capacity; make them, and learn about length, shape and technology; cover them, and learn about area; undo them, and learn about shape; and study them to enhance their awareness of patterns. The activities in this chapter use paper bags as a vehicle to develop a range of mathematical ideas.

Key vocabulary

full, empty, half full, nearly full, larger, smaller, square, triangle, rectangle, circle, about the same size as, longer than, shorter than, wider than

Make your own bag

Group size:

up to six children.

You will need:

A4 paper; scissors; glue.

The activity

Explain to the children that they are going to make their very own paper bag, maybe to keep something special in. Give them a sheet of paper each and ask them to think about how they could use it to make a paper bag. Accept and discuss any suggestions from the children and let them try them out.

The children may need two or three tries to make a bag that they are happy with, so have plenty of spare sheets of paper available. Their suggestions might include folding, making a cone, or cutting and sticking. As the children work, discuss the techniques they are using and the shapes they are making.

Let the children decorate and make handles for their bags.

Extension and differentiation

- To offer additional support, give each child a paper bag and help them to take it apart carefully to see how it is made. Is the sheet of paper much bigger than they expected?

- Give the children different-shaped paper (circles, triangles, squares, rectangles, etc.). Which shape makes the best bags?

- Set them the challenge of making a bag that is the right size to hold something specific, such as a teddy.

Sort

Group size:

up to six children.

You will need:

a collection of different-sized and patterned paper bags.

The activity

Show the bags to the children and talk about their shape and size ('Are the bags all different sizes?' 'Can you find two the same?' 'Which one is the largest?' 'What letters can you see on the bags?' 'How are they fixed together?').

Decide on one aspect that interests the children and use this as a criterion for sorting. This might be all the bags that are squares, or all those that are bigger than a hand. Record the results on a sorting diagram.

Extension and differentiation

- Sort the bags according to a range of other criteria: how long they are, how many cotton wool balls they hold, their colour, etc.

- Instead of using commercially produced bags, use those that the children have made.

- Provide the children with a set of toys, and ask them to put one toy in each bag, trying to match the size of the toy to the size of the bag.

Sorting paper bags

Cotton wool balls

Group size:
up to six children.

You will need:
a collection of small paper bags of different sizes; cotton wool balls.

The activity
Show the bags to the children and talk about their shapes and sizes.

Ask the children each to choose a bag, and fill it with cotton wool balls. Compare the number of balls in each bag, either by lining up the balls and finding out who has the longest line, or by counting them. Discuss the outcomes ('Were there two bags that held the same number of balls?' 'Whose bag held the most cotton wool balls?' 'Whose bag held the least?').

Extension and differentiation
- To reinforce the activity, try filling the bags with other items, such as interlocking cubes or ping pong balls. How do they compare with the cotton wool balls in terms of capacity?

- Let the children choose three bags each and fill them with cotton wool balls. Help them to write the appropriate number on each bag to show how many balls it holds. Ask the children to put the bags in order of how many balls they hold.

Dressing-up bags

Group size:
up to six children.

You will need:
a collection of paper bags of different sizes.

The activity
Explain to the children that they are going to dress up in paper bags.

Place all the bags in the centre of the group. Let the children take turns to pick a bag and suggest how it might be worn – on their head, foot, leg, hand, finger, etc. Explain that for them to be able to keep the bag, it must fit on the body part they have decided on, and not fall off. Then let them try them on. The first player to successfully wear five bags without them falling off wins the game!

Emphasise to the children that they should not play this game on their own and should never use plastic bags.

Extension and differentiation
- Change the rules of the game so that to win they need to wear more or fewer bags.

- Have the children make the bags for the game in advance. They will need to make a range of bags which will fit over different parts of their bodies.

Assessment

The assessment of the skills addressed in this chapter are covered more fully in the chapters *Introducing shape* (pages 4–7) and *How long? How tall?* (pages 32–35). In addition, watching and listening to children as they take part in these activities provides a good environment for identifying:

● whether children have made a connection between shapes and the number of sides they have, and how well they use of the language of measurement.

● whether skills addressed in other units have been remembered and used, and whether or not they need experience of further activities to extend their understanding. For example, if a child has difficulty finding bags to fit parts of their bodies in *Dressing-up bags*, it may be useful for them to try some of the activities in the chapter *Comparing* (pages 28–31).

Evidence of the children's learning

The children enjoyed wearing the paper bags, and soon were able to choose appropriately sized bags for hands and feet. One child took the activity home and came back the next day with a bag big enough for her to get right inside. This prompted others to bring in paper carrier bags that they could get both legs in or 'a whole head and my arms'!

The children got very involved in making bags, and most made more than one, trying to make them in different shapes with each attempt.

Other quick ideas

● Try to copy some of the patterns on the bags.

● Undo bags along their joins to use as templates to make new ones.

● Use a bag to make a parachute, by attaching cotton to it.

Involving parents

Ask parents to help children find bags at home and to compare their sizes. They could also help the children to make bags with handles to carry something special – a teddy? a book?

Glossary

Angle

A unit of turn between two straight lines or planes.

Capacity

The amount of space inside something.

Circle

The shape made by a line which has all its points the same distance from a single central point.

Conservation of measure

The fact that a measure stays the same whether or not the shape or layout is changed (for example, a piece of string is the same length whether it is held straight or otherwise).

Displacement

The increased water level caused by placing an object in a container of water. Used to measure the volume of irregular objects.

Equilateral triangle

A regular triangle having three sides and angles the same.

Irregular shape

A shape which does not have all its sides and angles the same.

Kite

A quadrilateral with two pairs of adjacent sides which are equal in length.

Parallelogram

A quadrilateral with two pairs of opposite parallel sides.

Polygon

A two-dimensional closed shape with straight sides.

Polyhedron (plural polyhedra)

A three-dimensional closed shape with straight sides.

Quadrilateral

A polygon with four straight sides.

Rectangle

A quadrilateral with two pairs of opposite parallel sides and right-angled corners.

Regular shape

A shape which has all its sides and angles the same.

Rhombus

A quadrilateral which has all four sides the same length.

Square

A quadrilateral with four equal sides and right-angled corners.

Tessellate

To join identical shapes by tiling, leaving no spaces at all.

Triangle

A polygon with three sides.

Vertices

The corners or points of a polygon.

Volume

The amount of space something takes up.

First published 2000 by A & C Black (Publishers) Ltd, 35 Bedford Row, London WC1R 4JH

The author and publisher wish to thank the staff and children of Knowles Nursery and Knowles First School, Bletchley, Milton Keynes, for their help in the preparation of this book.
ISBN 0-7136-4927-5
A CIP catalogue record for this book is available from the British Library.

Printed in China through Colorcraft Ltd.